Invertebrate Animals as Indicators of Acidity in Upland Streams

G.S. Patterson[a]

Wildlife and Conservation Research Branch
Northern Research Station
Roslin, Midlothian EH25 9SY

[a]*Present address:*
Forestry Practice Division
The Forestry Authority
231 Corstorphine Road
Edinburgh EH12 7AT

and

B.R.S. Morrison

Freshwater Fisheries Laboratory
The Scottish Office
Agriculture and Fisheries Department, Faskally
Pitlochry, Perthshire PH16 5LB

LONDON: HMSO

ISBN 0 11 710317 9
FDC 116.91: 142 (410)

KEYWORDS: Acid waters, Indicators,
Invertebrates, Upland streams

Enquiries relating to this publication
should be addressed to:
The Technical Publications Officer,
The Forestry Authority, Forest Research Station,
Alice Holt Lodge, Wrecclesham,
Farnham, Surrey, GU10 4LH.

Printed in the United Kingdom for HMSO
Dd 0294465 C22 4/93

Front cover: Kick sampling for invertebrate animals in a forest stream at Cardrona, the River Tweed catchment, Borders Region.

Contents

Invertebrate Animals as Indicators of Acidity in Upland Streams

Summary

Studies in the UK and elsewhere have shown that acid freshwater habitats have different floras and faunas and fewer species in most taxonomic groups, when compared with near-neutral waters. The presence or absence of certain common species that are sensitive to acid waters can be used to assess the prevailing chemistry of a water body. The method described in this Field Book enables managers to do this for forest streams in upland areas using a limited number of readily recognised invertebrate animals. It may also be used for streams outside woodland. The method is suitable for qualitative monitoring over a number of years to show trends in the ecological status of selected streams.

Invertébrés comme Indicateurs de l'Acidité des Ruisseaux en Altitude

Résumé

Des études réalisées au Royaume-Uni et dans le reste du monde ont démontré que les habitats d'eau douce acide possèdent une flore et une faune différentes et moins variées dans la plupart des groupes taxonomiques par rapport aux habitats d'eau approximativement neutre. La présence ou l'absence de certaines espèces sensibles aux eaux acides peut être employée pour déterminer la composition chimique prédominante d'un cours d'eau. La méthode décrite dans ce rapport d'étude permet aux techniciens des eaux et forêts d'analyser les ruisseaux forestiers en altitude grâce à un petit nombre d'invertébrés facilement reconnaissables. Elle peut également être employée dans les ruisseaux non forestiers. Cette méthode permet ainsi d'effectuer une analyse qualitative sur plusieurs années afin de suivre l'évolution écologique de cours d'eau sélectionnés.

Wirbellose Tiere als Indikatoren des Säuregrades von Wasserläufen im Hochland

Zusammenfassung

Untersuchungen in GB und anderswo haben gezeigt, daß saure Süßwasserlebensräume im Vergleich zu normalen Gewässern andere Floras und Faunas und in den meisten Tierordnungen weniger Arten besitzen. Die An- oder Abwesentheit bestimmter alltäglicher Arten, welche saurem Wasser gegenüber empfindlich sind, kann benutzt werden, um die vorherrschende Chemie eines Gewässer zu bewerten. Die Methode, die in diesem Feldbuch beschrieben wird, ermöglicht es Forstverwaltern, dies für Waldbäche im Hochland, anhand von einer begrenzten Anzahl, leicht erkennbarer wirbellosen Tiere zu tun. Sie kann auch für Wasserläufe außerhalb des Waldes benutzt werden. Die Methode eignet sich zur qualitativen Überwachung über einen Zeitraum von mehreren Jahren, um Tendenzen im ecologischem Status ausgewählter Bäche aufzuzeigen.

Introduction

Much research in recent years has focused upon acidification of surface waters in the UK and elsewhere. This has shown that streams, lochs and lakes in some upland areas on poorly buffered lithologies are currently acidic to an extent that significantly affects their ecology (Figure 1), and large areas of upland Britain are classified as susceptible to acidification because of their geology and soils.

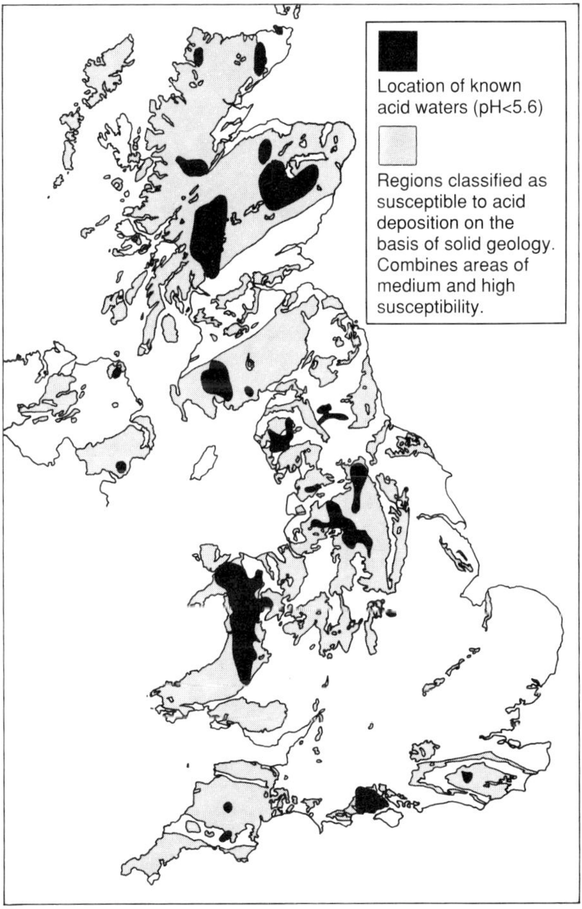

Figure 1 Distribution of acid waters and areas susceptible to acidification in the UK.
(Adapted from UK Acid Waters Review Group, *Acidity in United Kingdom fresh waters,* HMSO, London, 1989, map on page 4.)

Acidification is to some extent a natural phenomenon but the rate of acidification has been increased during the past century by the deposition of oxides of sulphur and nitrogen produced by combustion of fossil fuels (UK Acid Waters Review Group, 1989).

Some studies have shown that in areas with poorly buffered soils and high levels of atmospheric pollutants there is an association between the amount of closed-canopy forest in a catchment and the levels of acidity of streamwater (e.g. Harriman and Morrison, 1982; Ormerod *et al.*, 1989).

Forest canopies enhance the capture of pollutants from the atmosphere and can thereby increase the risk of acidification leading to biological effects in areas with a combination of high atmospheric deposition, high rainfall and low acid neutralising capacity. The main areas at risk are parts of Central and South-west Scotland, South-west Cumbria, the Pennines and Central and North Wales (Department of Environment, 1990; Nisbet, 1990).

The planned reduction of 60% in emissions of sulphur dioxide in the UK by 2003 compared with 1980 levels (Department of Environment, 1990) is expected to lead to a reduction in streamwater acidity over several decades. The Forestry Commission's *Forests and water guidelines* (Forestry Commission, 1992) take account of this and of calculations of the critical load of pollutants which individual areas can tolerate. They also contain guidance on other measures intended to ameliorate surface-water acidity in forests. These include restricting the amount of high altitude planting in sensitive catchments, applying lime, and the development of herbaceous vegetation and broadleaved woodland in riparian areas.

This Field Book provides foresters and other land managers with a simple method of assessing and monitoring their streams in terms of water chemistry and ecological status.

The biological effects of acidity

Acidity is measured as pH, a logarithmic scale based on the reciprocal of the concentration of the hydrogen ion (H^+) in the water. Water at pH 7 is chemically neutral. Values below 7 indicate acidity; values above, alkalinity. Acid water, particularly at pH < 5.6, can upset the natural salt balance in many species of fish and invertebrates and if the acid conditions persist they may prove fatal (Harriman *et al.*, 1987). When the mean pH falls below about 5.6 a shift in species composition and a net decline in species diversity starts to occur in most taxonomic groups in both standing and running waters (UK Acid Waters Review Group, 1989). The main changes under these conditions are summarised in Table 1.

Other aspects of water chemistry related to acidity also have an important influence on plants and animals. Aluminium and heavy metals such as zinc, copper and cadmium can be toxic when present as free ions. These ions become more soluble as acidity increases (i.e. as the pH value decreases). When calcium concentrations are low (often < 2 mg l^{-1} in acid waters) there may not be enough calcium for the needs of some animals, e.g. for the shells of snails and mussels, and the hard 'shell' or exoskeleton of shrimps. Calcium and magnesium are normally present as bicarbonates which are important as buffers since they reduce the levels of free hydrogen (H^+) and heavy metal ions. Bicarbonates can exist in water with pH values as low as 5.5 and consequently some buffering is possible even under these conditions. So-called 'harder' waters, where the total calcium and magnesium concentrations are equivalent to at least 10 mg l^{-1} of calcium carbonate, are unlikely to become significantly acidic and could support a wide range of species.

The presence of organic material such as humic acids in the water can also mitigate the harmful effects of aluminium on animals, especially fish, by combining with dissolved aluminium to form non-toxic complexes.

It is therefore hard to predict the detailed effects of a given pH level upon wildlife without knowledge of these other factors. The damage to sensitive wildlife is greatest when the pH values are lowest, which tends to be when aluminium and heavy metal con-

Table 1. Main types of effects of increasing acidity[a] upon various wildlife groups (summarised from UK Acid Waters Review Group, 1989)

Group	Main effects
Micro-organisms	Bacterial and fungal activity is reduced and decomposition slows down. Diversity of diatoms reduced.
Plankton	Species composition changes for phyto- and zooplankton. Species-richness declines for zooplankton, e.g. *Daphnia* water fleas, at pH < 5.5 in lakes.
Algae	Changes in species composition towards filamentous algae.
Higher plants	Change in community composition in streams towards acid-tolerant bryophytes such as the liverwort *Scapania undulata*. In larger rivers and lakes the macrophyte flora is reduced in diversity.
Invertebrate fauna	Fewer species present and increasing dominance by a few taxa, especially stoneflies in streams. Sensitive taxa, e.g. crustaceans, molluscs, some beetles and caddis flies and most mayflies, are gradually lost. Productivity sometimes declines.
Fish	Salmonid fish gradually become scarce or die out below pH 5.6. Brown trout are less vulnerable than salmon or rainbow trout. Non-lethal effects such as growth reduction and reproductive failure occur first.
Amphibians	Spawning, metamorphosis and growth affected, mainly below pH 5.
Other higher animals	Dippers become scarce: mayfly and caddis fly larvae are major food sources. Fish-eating birds may decline in numbers. Otters could become scarce in acid areas but there is no evidence to show this to date.

[a] Defined as mean pH falling below 5.6.

centrations are highest. Water chemistry can change rapidly during spates when the pH can drop from 6.5 to 5.0 within a few hours in poorly buffered head-water streams over acid rocks.

Assessing the acidity of forest streams

A forest manager may have several reasons for wishing to assess streams in terms of acidity and its effects on stream ecology. The main questions are likely to be:

1. Which streams are acidic and are inclined to have a low diversity of animals and plants? In particular which streams are likely to be unsuitable as spawning and nursery streams for salmon and trout?
2. Which streams are not acidified and have a water chemistry suitable for a wide range of species, including fish?
3. What is the trend of acidity in particular streams?

The answers to these questions are clearly important to staff who manage streams and riparian habitats for wildlife conservation in general, or for fisheries in particular. They would enable managers to target expenditure on conservation or fisheries upon un-acidified streams which have higher potential, variety and productivity. On the other hand, they would also assist in identifying streams where the amelioration of acid conditions could be considered.

Invertebrate animals as indicators of water quality

When assessing water quality in streams, sampling the invertebrate fauna has certain advantages over direct measurements of pH or ion concentrations. The range of species present reflects living conditions over a period of months or years, while a chemical analysis is only valid for the time the sample was taken. As has been stated, the crucial chemical parameters associated with the harmful effects of acidity, pH, calcium and aluminium concentrations can vary considerably within a period of hours or days. Measurements must therefore be made over a considerable period and in all flow conditions to obtain a reliable picture of water chemistry. Many aquatic invertebrates have fairly specific habitat requirements and thrive within a limited range of water chemistry. Such 'indicator species' can be used to assess the status of the water in broad terms, i.e. whether it is acid or alkaline, and how well buffered it is. From that assessment, its suitability for fish or for the conservation of other species can be determined.

There is no well-defined limit of pH or calcium concentration below which acid-sensitive animals are never found. Environmental influences, and variations in the genetic make-up of individual organisms, mean that specimens of acid-sensitive species may be collected occasionally from habitats that might be thought unsuitable. Such habitats are, however, likely to be borderline cases with respect to acidity and in any one sample the number of sensitive animals caught is normally limited to one or two. This contrasts with much greater populations in less acid waters.

Indicators of acid streams

A wide range of studies in the UK, Scandinavia and elsewhere has shown that a number of common invertebrate animals are sensitive to acid water conditions at or close to a mean pH level of 5.6. Crustaceans such as *Gammarus pulex* (Plate 1) and molluscs, e.g. *Ancylus fluviatilis* (Figure 2), need calcium for their hard outer skeletons or shells and are absent in soft acid waters with mean total hardness $< 10–15$ mg l^{-1} of $CaCO_3$. Most mayfly larvae (Ephemeroptera, Figure 3, Plates 2 and 3) and some caddis fly larvae (*Hydropsyche* species, Plate 4) are also scarce or absent when the mean pH falls below about 5.7. Many stoneflies (Plecoptera, Figure 4, Plate 5), on the other hand, can tolerate a mean pH down to 4 or less and usually dominate the fauna of acid streams, although they are also found in non-acid streams.

It should be borne in mind that many other factors affect the distribution and abundance of the indicator species, for example, the composition of the streambed, the flow regime and also the type of riparian vegetation, which influences the food supply and light and temperature conditions. The *absence* of acid-sensitive indicator species cannot therefore be taken as conclusive evidence of unsuitable water chemistry.

However their presence can give a good indication of suitable water chemistry conditions for other animals and plants, as Table 1 suggests.

Suitability of the water for fish is likely to be of particular interest to managers. In general the relationship between the tolerance of most acid-sensitive invertebrates and that of salmonid fish is fairly close, although brown trout can survive slightly more acid conditions than some of the indicator invertebrates. However even where trout are present in acid waters of pH 4.5–5.5, where the sensitive invertebrates such as mayflies may be absent, the population is likely to be under stress, as demonstrated by low population numbers (e.g. Turnpenny *et al.*, 1987) and frequent reproductive failure because of the vulnerability of eggs and fry to acid conditions. Salmon and rainbow trout are more sensitive than brown trout. Pike and perch show intermediate sensitivity while eels are quite acid tolerant, sometimes being found below pH 4.5.

The invertebrate indicators proposed here are therefore a reasonable guide to likely suitability of water chemistry for most fish species. The species proposed for use as indicators are all common in upland streams and can be recognised readily with the aid of a hand lens (Table 2).

Table 2. Acid-intolerant indicators

a. *Group 1*: Animals normally absent if mean pH < 6

Species	Normal minimum mean pH	Behaviour
Gammarus pulex (freshwater shrimp)	≥6.0	Between stones feeding on plant litter
Glossosoma species and Agapetus species (caddis fly)	6.0	Grazing insects inside small stony cases attached to rocks
Ancylus fluviatilis (freshwater limpet)	6.0 Occasionally down to 5.5 if calcium levels are high	Graze on algae on stones
Limnaea peregra (wandering snail)	6.0 Occasionally down to 5.5 if calcium levels are high	Graze on algae on stones
Asellus aquaticus (freshwater slater)	6.0 Occasionally down to 5.5 if calcium levels are high	Found on silty or sandy bed with decaying plant material

b. *Group 2*: Animals normally absent if mean pH < 5.5

Species	Normal minimum mean pH	Behaviour
Hydropsyche species (caddis fly)	5.5 - 6.0	Spin silken traps among stones to capture food items
Ephemeroptera (mayflies):		
Baetis species (mainly *Baetis rhodani*) round-bodied nymphs	5.5 Occasionally found at pH 5.2	Swim rapidly over stony bed, feeding on detritus and algae
Family Heptageniidae: flat-bodied mayfly nymphs (e.g. *Rhithrogena semicolorata*)	5.5 Occasionally found at pH 5.2	Crawl over stones feeding on detritus

1

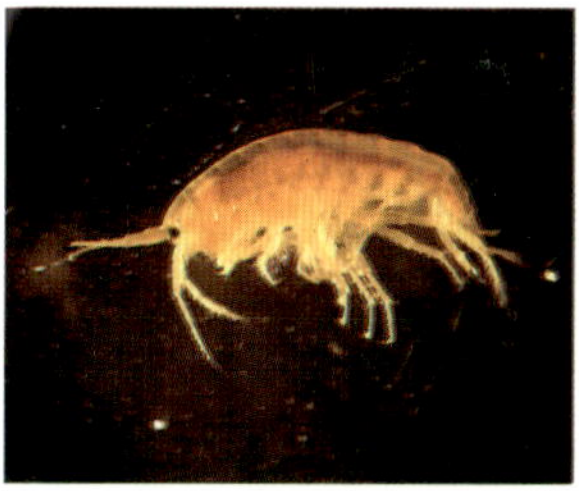

5

Plates 1–4. Acid-intolerant indicators.

1. The freshwater shrimp (*Gammarus pulex*): an acid-intolerant species normally absent where the mean pH is under 6. (B. Morrison)

2. A flat-bodied *Ecdyonurus* mayfly nymph (family Heptageniidae). (D. Morrison)

3. *Baetis rhodani* mayfly, a very common species in stony streams. (B. Morrison)

4. *Hydropsyche* species, a caddis fly larva, which requires a mean pH of at least 5.5. (B. Morrison)

5. Most stoneflies (order Plecoptera) can tolerate acid streams and are normally the dominant invertebrate animals where mean pH falls below about 5. (Institute of Freshwater Fisheries)

Plate 6. Excessive shade inhibits moss and algal growth in streams and the associated invertebrate fauna. Do not sample shaded sections. (40013)

Plate 7. Allow 3 years for stream invertebrates to adjust to bankside felling before sampling. Plant growth on the streambed and banks provides increased food and shelter. Cut branches should be kept away from streams (Forestry Commission, 1992) but if *small* amounts do reach the channel during later floods, any effect on invertebrates should be local and short-lived. (G. S. Patterson)

Plate 8. top A well-lit stony streambed is ideal for sampling. (B. Morrison)

Plate 9. Streamside clearance and planting of broadleaved trees in tree shelters. Indicator species sampling can assist in allocating priorities for expenditure on environmental improvements like this. (39764)

Crustaceans

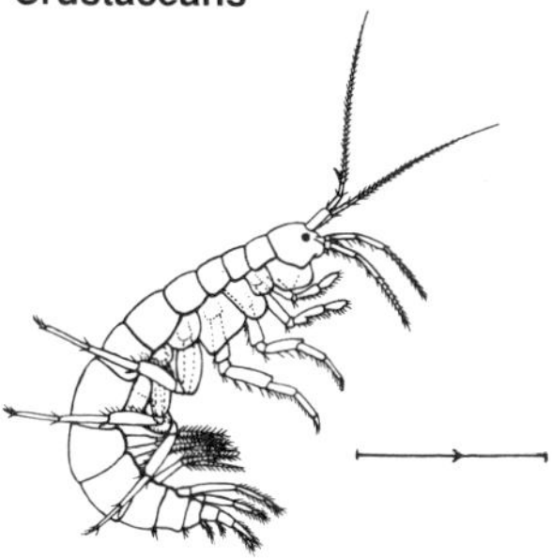

Gammarus pulex:
freshwater shrimp

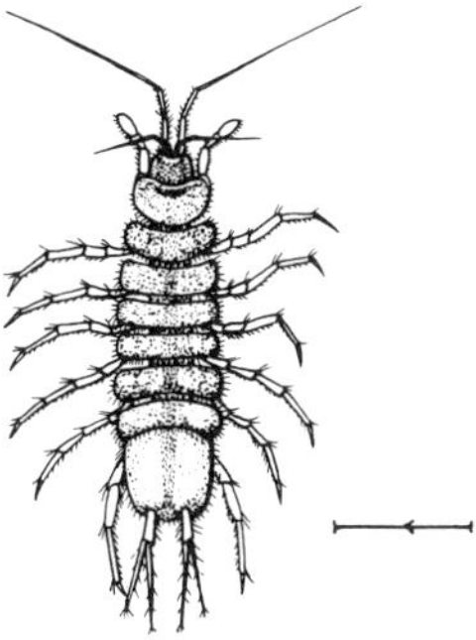

Asellus aquaticus:
freshwater slater

Molluscs

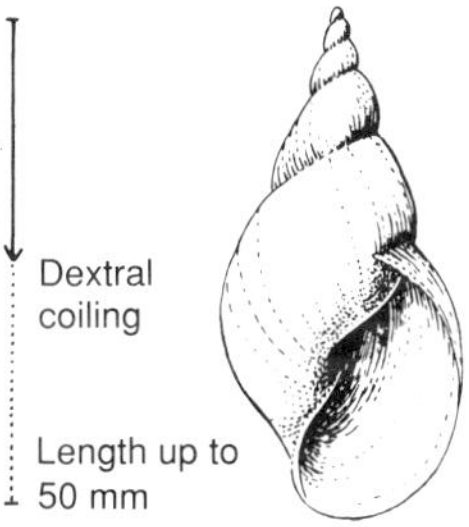

Limnaea (includes *Limnaea peregra,* the wandering snail)

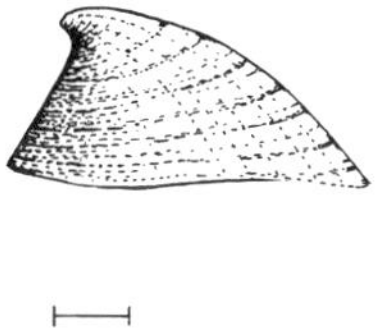

Ancylus fluviatilis:
freshwater limpet

Insects
Caddis fly larvae

side
view

upper
view

Glossosoma and *Agapetus* species : cases

Figure 2 Acid-intolerant indicators: Group 1.
Bars show actual length of animals.
(Adapted from Croft, *A key to the major freshwater invertebrates,* Field Studies Council, 1986.)

Insects

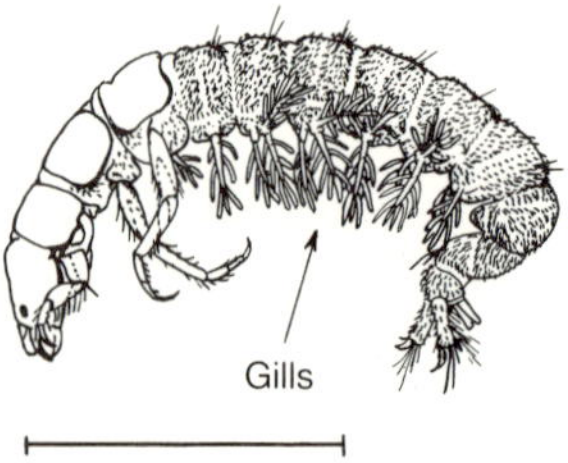

Caddis fly: *Hydropsyche* species

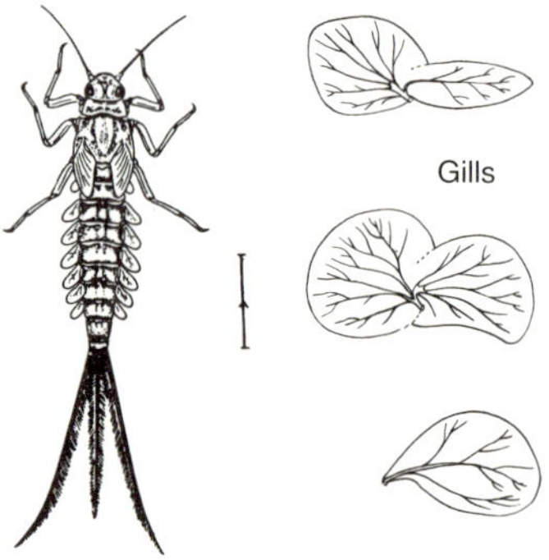

Round bodies
Middle tail shorter than other two
May have black rings on tails
No spine on last body segment
Gills are plate-like

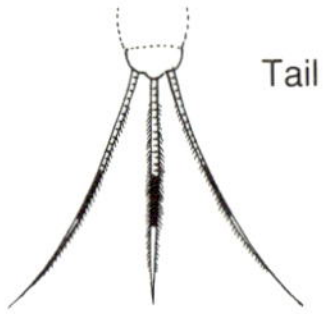

Mayflies: *Baetis* species
(Baetis rhodani, Baetis muticus)

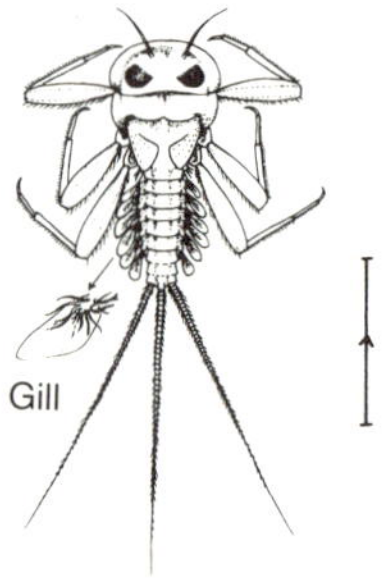

Flat bodies

Three tails
Middle tail longer than others

Mayflies: Family Heptageniidae
(e.g. *Rhithrogena semicolorata)*

Figure 3 Acid-intolerant indicators: Group 2.
(Adapted from Croft, *A key to the major freshwater invertebrates,* Field Studies Council, 1986.)

Insects

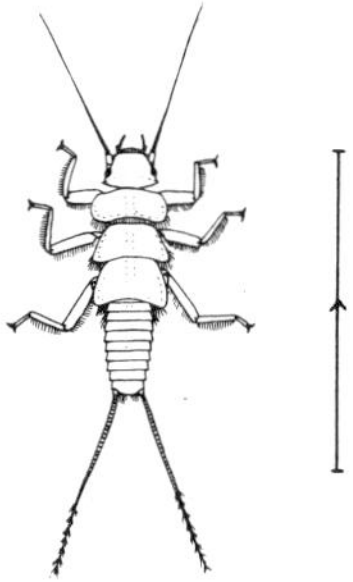

Length
up to
50 mm

Two
tails

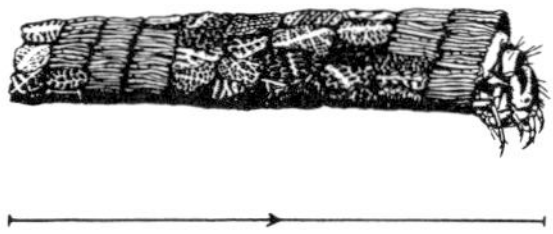

Note: most but not all of these two orders are acid-tolerant

Plecoptera: stonefly nymphs

Trichoptera: caddis fly nymphs

Mayfly nymphs

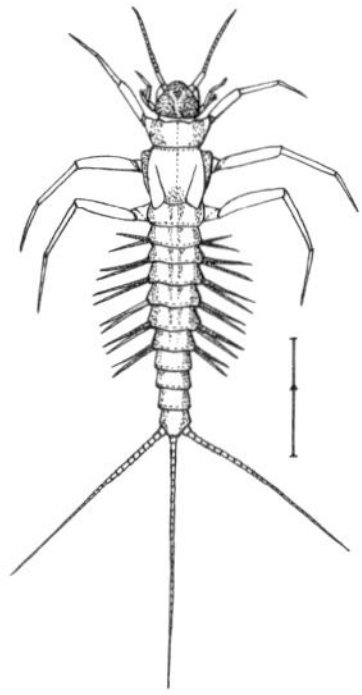

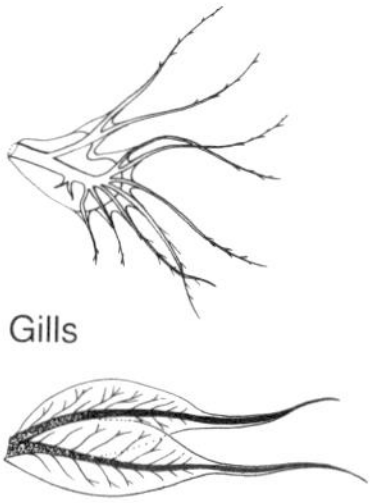

Gills

Spines but no
black bands
on tails
Long tails
Gills forked/branched

Leptophlebia species

Round body
Tails of *equal* length
Black bands on tails
(but not black rings)
and a spine on the
last body segment

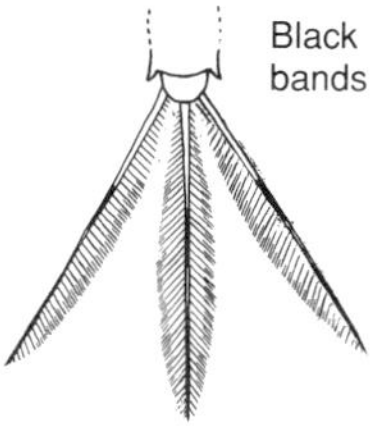

Black
bands

Siphlonurus species

Avoid confusing these with acid-intolerant mayflies (Figure 3)

Dytiscidae: swimming beetles
Hydroporus species

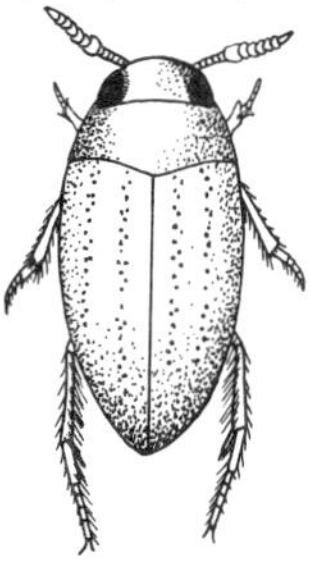

adult

Often black-yellow
Rapid swimmer

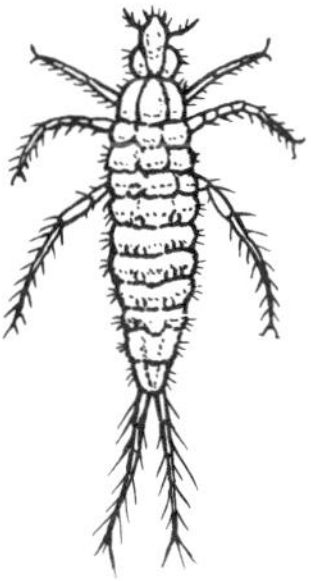

Short
antennae

larva

Figure 4 Acid-tolerant groups.
(Adapted from Croft, *A key to the major freshwater invertebrates,* Field Studies Council, 1986.)

Sampling method

The method assumes sampling in a standardised manner. Sampling should be done in April so that most of the animals present will have lived through the autumn and winter period when the most acid episodes occur. Most insect larvae or nymphs present will have hatched the previous summer or autumn, and will be well developed by April and easier to identify. Sampling after April is not suitable as many overwintered insect larvae will have become adults and left the water.

Selection of sampling sites

The sampling method must be consistent to enable a comparison to be made between sites and between years. To reduce the chances of other factors masking the influence of water chemistry in determining the distribution of the indicator species the sampling sites should be:

- Perennial streams of at least 1 m in width.
- Unshaded by closed canopy plantation, and not clearfelled within the last 3 years (Plates 6, 7, 8). Sections of open streamside of at least 100 m next to closed-canopy sections may be suitable, however.
- Unaffected by recent siltation or large scale bank erosion.

The sampling sites should have a streambed which is a suitable habitat for most of the indicator species. It should be stony, variable in composition, with some grit and sand (Plate 8), but with over 50 % of the bed consisting of stones or cobbles of up to 25 cm diameter. Larger boulders of 30–50 cm diameter are often present in such streams. Streams with predominantly silty or bouldery beds have a limited range of very specialised animals.

The number and location of sites selected will depend upon the resources available and the manager's objectives. The main reasons for sampling are likely to be:

1. Descriptive survey of the streams in a forest block.
2. Monitoring, where representative sections of streams would be sampled repeatedly.

For a descriptive survey the basic requirement is one sample site per stream. If resources allow, then additional samples in well-spaced sections of each stream, preferably at least 200 m apart, are desirable to reduce the likelihood of local physical or biological factors masking the influence of water chemistry.

For monitoring, one site per sampled stream should suffice, given sufficient sampling occasions (at least five) to allow for year to year variability in physical and chemical conditions.

Sampling procedure

The equipment required comprises:

- Hand net with 20 × 20 cm or 25 × 25 cm opening and mesh size between 0.5 and 1.0 mm (Figure 5): too fine a mesh will trap large amounts of detritus and silt. Handle length should be at least 50 cm (see Figure 5). Suitable nets may be obtained from suppliers of equipment for aquaria and fish farms.
- White tray approximately 30 × 20 cm, e.g. a photographic tray.
- Hand lens with magnification × 8 or × 10.
- Several wide-mouthed jars or similar containers, approximately 500 ml capacity, with lids.
- Preservative (made up of 7 parts industrial alcohol to 3 parts water). The addition of a small amount of glycerol, about 2 %, reduces the risk of specimens drying out due to evaporation of the alcohol.
- Forceps or tweezers and a saucer or similar dish for closer examination of individual specimens with the hand lens.

At each sample point place the bottom edge of the net on the bed of the stream close to the centre of the channel. Kick into the stones just upstream of the net while walking slowly upstream, for a period of 2 minutes, allowing the disturbed material to drift downstream and be caught in the net. The samples should be transferred from the net to a wide-mouthed container. If they cannot be examined within 2–3 hours they should be covered with preservative and the container closed. During the examination the sample should be transferred to a white tray and any plant material teased apart and the animals lifted out with forceps into a saucer or other small dish. Most animals can be identified without the hand lens using the diagrams in Figures 2, 3 and 4. Identification of the *Baetis* mayflies (normally *Baetis rhodani*) may require the sample to be preserved and examined later with a hand lens. The Further Reading lists sources of further information and descriptions of a wider range of invertebrates of upland streams.

Recording

Details of the sample site should be entered on the Site Description Sheet (Appendix 1). This will help in the interpretation of sample results, especially where monitoring is intended, because it will reveal changes in substrate type and riparian vegetation over a period of years. The substrate should be recorded for a 1 m strip over the length of streambed which is sampled (approximately 20 m).

Only the indicator animals listed in Tables 2 and 3 should be recorded. Presence or absence of species in each sample should be recorded on the Indicator Record Sheet shown in Appendix 2. Where repeated visits are intended for monitoring, the sample location should also be marked by a stake on the bank above the spate zone. Location of samples should be entered on the data sheet.

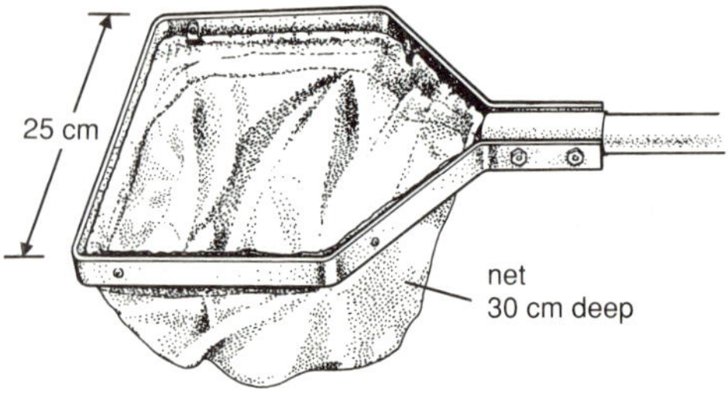

Net composed of 1mm mesh knitted polyester with nylon collar

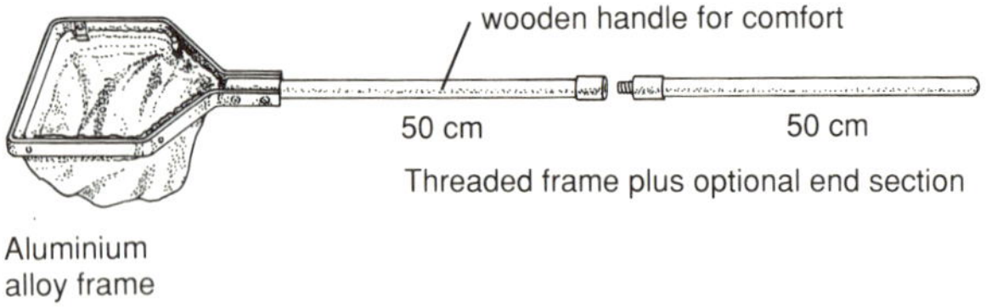

Figure 5 Specification of sampling net.
(This specification is supplied by GB Nets, Linden Mill, Hebden Bridge, West Yorkshire, HX7 7DP.)

Table 3. Acid-tolerant groups

Species	Normal minimum mean pH	Behaviour
Order Plecoptera (stonefly nymphs) many but not all species are acid-tolerant	≤ 4	Crawl over stones. Larger species predatory, smaller ones feed on detritus
Order Trichoptera (caddis) many species, with or without cases. Many but not all species are acid-tolerant	approx. 4	In stony streams. Cased caddis generally feed on detritus or graze on algae; free living caddis are often carnivorous
Family Leptophlebiidae (mayfly nymphs)	approx. 4	As for *Baetis rhodani:* swim rapidly over stony bed, feeding on detritus and algae
Family Siphlonuridae (mayfly nymphs)	approx. 4	As for *Baetis rhodani:* swim rapidly over stony bed, feeding on detritus and algae
Family Dytiscidae (swimming beetles)	approx. 4	Swim rapidly when disturbed. In stony or silty streams. Many species carnivorous

Monitoring

Repeated samples can be used to build up a picture of the trend in water chemistry of a given stream or catchment. It is preferable to use the same sampling site on each occasion to reduce variability due to site factors. However if the sampling site is drastically altered in its substrate composition it may be better to find a suitable site on the same section of stream as close as possible to the original one.

Where monitoring is intended the number of individuals could be recorded on the record sheet to add to the information about trends over time. To reduce recording effort large numbers could be recorded as 'over 20' or another appropriate threshold value.

Interpreting the results

The decision chart in Figure 6 should be used with the completed Indicator Record Sheet (Appendix 2).

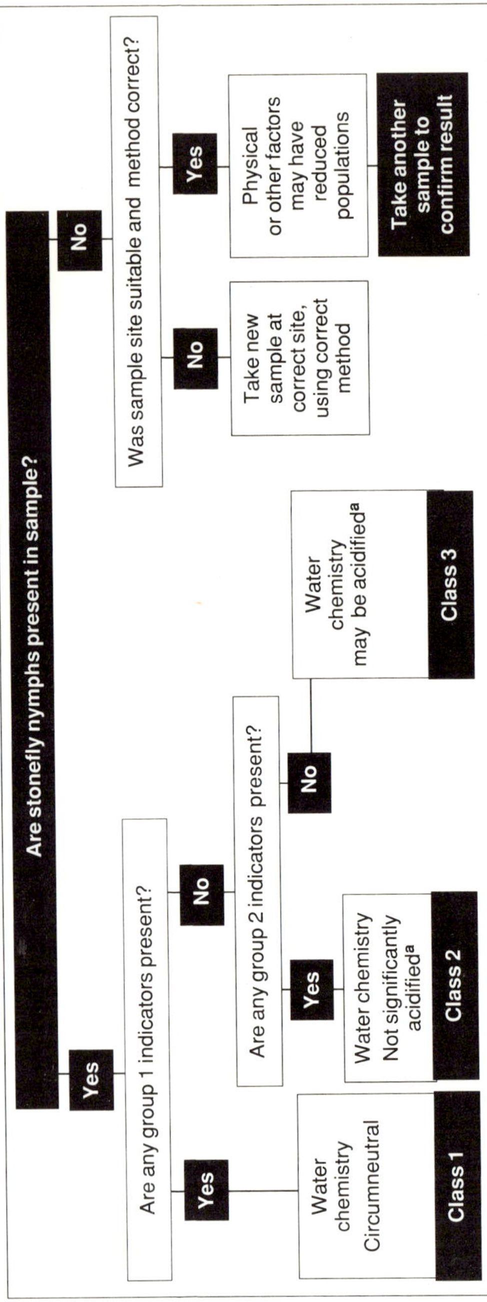

[a] Further samples at the same site and others on the same stream may be considered to increase the confidence of the diagnosis, where the number of group 2 indicator species present in the first sample is 0 or 1.

Figure 6 Interpreting samples: decision chart to be used alongside Indicator Record Sheet (Appendix 2).

Classifying the water chemistry status

The definition and ecological significance of the three classes is important.

Class 1

Streams with a water chemistry suitable for the great majority of plants and animals. They are sufficiently alkaline to be buffered against most acid spate waters and pH is unlikely to drop below 5.6 with the mean probably exceeding 6.0. Salmonid fish would not suffer any significant stress from water chemistry.

Class 2

Streams which are not significantly acidified. They could possibly be circumneutral like Class 1, but merely lack group 1 indicators through the chance location of the sample. Their mean pH is unlikely to be much less than 5.6 although where aluminium and heavy metal levels are low and/or organic content is high the mean could be down to about 5.3. These streams are likely to be suitable for most wildlife species except perhaps the most sensitive taxa.

The water chemistry is likely to be suitable for fish populations and aquatic birds including the dipper. These streams could be vulnerable to acidification in future however.

Class 3

Streams which may be acid to the point where wildlife is significantly affected across a wide range of groups (see Table 1).

These effects include the reduction of populations of salmonid fish, especially salmon, and reduced invertebrate animal diversity.

The diagnosis of Class 3 cannot be certain because lack of acid-sensitive indicators in samples could be due to sampling error or causes other than water chemistry. Taking further samples would help to eliminate these factors and increase certainty.

Chemical analysis of streams identified as Class 3 should be considered to further improve the diagnosis.

Management implications

Survey and monitoring of streams using this method will provide the forest manager with information on the status and trend of streams with respect to acidity and biological diversity. Often this may be the main purpose of the sampling programme, as it will enable the manager to judge how far his/her streams meet expected standards and objectives.

The information can also be used to help in drawing up management programmes. For example, it would help to identify sites with greater sensitivity to habitat damage (these would normally be Class 1 streams which are likely to be the most diverse). It would also help to identify sites with the greatest potential return on investment made to improve wildlife conservation or recreation values (Plate 9). These would normally be Class 1 and 2 streams where acidity would not be a constraint upon their biological diversity. Class 1 and 2 streams might be made a greater priority for removing excessive shade by conifers, for example.

Management to improve the habitat structure for fish or dippers, or possibly otters, might also be directed to Class 1 and 2 streams. On the other hand, Class 3 streams could be selected for amelioration of acidity as appropriate methods become available.

Conclusion

This Field Book presents a method which can be used by rangers and similar staff to assess stream chemistry using a limited number of indicator animals that can be readily identified by the naked eye or hand lens. The method can be used for survey and monitoring purposes and can thereby assist managers to determine whether their streams are satisfactory in terms of water chemistry, and to select appropriate management regimes.

ACKNOWLEDGEMENTS

The authors would like to thank Dr S.J. Ormerod, Mr C.M.A. Taylor and Dr T. Nisbet for their valuable comments on various drafts and the typing staff at Northern Research Station for their hard work in producing them. Thanks are also due to Field Studies Council Publications for their help and permission to reproduce the diagrams of invertebrate animals from P.S. Croft, *A key to the major freshwater invertebrates.*

REFERENCES

Department of Environment 1990. *This common inheritance — Britain's environmental strategy.* White Paper. HMSO, London.

Department of Environment and Forestry Commission 1991. *Forests and surface water acidification.* Department of Environment, London.

Forestry Commission 1992. *Forests and water guidelines,* 2nd edition. HMSO, London.

Harriman, R. and Morrison, B.R.S. 1982. Ecology of streams draining forested and non-forested catchments in an area subject to acid precipitation. *Hydrobiologia* **88**, 251–263.

Harriman, R., Morrison, B.R.S., Caines, L.A., Collen, P. and Watt A.J. 1987. Long-term changes in fish populations of acid streams and lochs in Galloway, South-west Scotland. *Water, Air and Soil Pollution* **32**, 99–112.

Nisbet, T.R. 1990. *Forests and surface water acidification.* Forestry Commission Bulletin 86. HMSO, London.

Ormerod, S.J., Donald, A.P. and Brown, J.J. 1989. The influence of plantation forestry on the pH and aluminium concentration of upland forest streams: A re-examination. *Environmental Pollution* **62**, 47–62.

Turnpenny, A.W.H., Sadler, K., Aston, R.J., Milner, A.G.P. and Lynam, S. 1987. The fish population of some streams in Wales and northern England in relation to acidity and associated factors. *Journal of Fish Biology* **31**, 415–434.

UK Acid Waters Review Group 1989. *Acidity in United Kingdom fresh waters.* Second Report to Department of the Environment. HMSO, London.

FURTHER READING

Croft, P.S. 1986. *A key to the major groups of British freshwater invertebrates.* AIDGAP Services, Field Studies Council of England and Wales, Shrewsbury.

Doughty, C.R. 1990. *Acidity in Scottish rivers. A chemical and biological baseline survey.* Scottish River Purification Boards/Department of Environment.

Fitter, R. and Manuel, R. 1986. *A field guide to the freshwater life in Britain and North West Europe.* Collins, London and Glasgow.

Macan, T.T. 1959. *A guide to freshwater invertebrate animals.* Longman, London.

Mellanby, K. 1975. *Animal life in freshwater*, 6th edition. Chapman and Hall, London.

Appendix 1 Site Description Sheet

Sample number ____________

Date ____________

Name of stream ____________ Location/Compt number ____

100 m Grid ref. ____________

Site details

Stream width (current wetted width, from water's edge to water's edge) ____________

Substrate. Percentage of streambed area that consists of the following substrate types (enter A, S, C, F or D):

Absent (A)	Scarce (S)	Common (C)	Frequent (F)	Dominant (D)
0 %	1–5 %	6–20 %	21–50 %	> 50 %

Bedrock	
Boulders > 25 cm	
Cobbles 6–25 cm	
Coarse and fine gravel 0.2–6 cm	
Sand, silt and clay < 0.2 cm	

Vegetation. Percentage of water surface for a distance of 30 m on each side of the sample point that is overhung (within < 0.5 m and > 0.5 m of surface, respectively) by the following vegetation types (enter A, S, C, F, or D):

Absent (A)	Scarce (S)	Common (C)	Frequent (F)	Dominant (D)
0 %	1–5 %	6–20 %	21–50 %	> 50 %

	< 0.5 m	> 0.5 m
Herbaceous vegetation (e.g. grasses and rushes)		
Deciduous woody vegetation (i.e. shrubs and trees)		
Coniferous trees		

Principal riparian land use, within 200 m upstream (tick box):

Moorland heathland	
Rough pasture	
Improved pasture	
Arable	
Deciduous woodland	
Coniferous woodland	
Urban	
Industrial	
Tip/waste	

Appendix 2 Indicator Record Sheet

Indicator group	Animal group/species	Presence[a]	Notes[b]
Acid-intolerant Group 1	*Gammarus pulex* (freshwater shrimp)		
	Glossosoma and *Agapetus* species		
	Ancylus fluviatilis (freshwater limpet)		
	Limnaea peregra (wandering snail)		
	Asellus aquaticus (freshwater slater)		
Acid-intolerant Group 2	*Hydropsyche* species (caddis fly)		
	Baetis species (mayflies)		
	Family Heptageniidae (flat bodied mayflies)		
Acid-tolerant	Plecoptera (stoneflies)		
	Trichoptera (caddis fly)[c] Leptophlebiidae (mayflies)[c] Siphlonuridae (mayflies)[c] Dytiscidae (swimming beetles)[c]		

Water chemistry category indicated (from Figure 6 ....

Notes

[a] Record presence for each taxon in sample.

[b] This column may be used to aid interpretation, e.g. records of abundance.

[c] These animals will not normally be required for diagnosing water chemistry. They may be helpful however in exceptional cases where stoneflies are absent in repeated samples (see Figure 6).